D1493568

THE LITTLE BOOK OF
ROSÉ

Published by OH!
20 Mortimer Street
London W1T 3JW

Text © 2021 OH!
Design © 2021 OH!

ISBN 978-1-80069-051-6

Compiled by: Lisa Dyer
Editorial: Stella Caldwell
Project manager: Russell Porter
Design: Tony Seddon
Production: Freencky Portas

A CIP catalogue record for this book is available from the British Library

Printed in Dubai

10 9 8 7 6 5 4 3 2 1

Images: Freepik.com

THE LITTLE BOOK OF
ROSÉ

SUMMER PERFECTION

CONTENTS

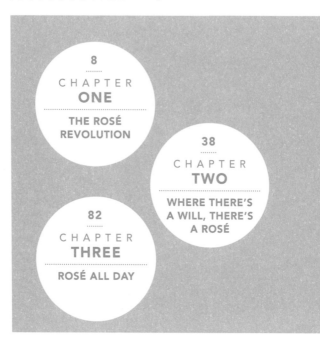

Introduction

With celebrities and influencers showing off their best life holding Insta-friendly glasses of rosé, you'd be forgiven for thinking the world doesn't spin without smiles, sunsets and this fabulously coloured beverage. Pale and delicate or robust and ruby red, rosé has captured the public's imagination and fuelled a revolution in the drinking world.

Although it is now enjoying an astronomical boost in popularity, rosé hasn't always been the wine of choice among the connoisseur and the cognoscenti. The ancients must've made some truly terrible potations on their way to refining winemaking, but we have to thank the French for putting the most excellent of elixirs on the map. From the first vines planted in the heartland of Provence to the rosé-filled carafes at sidewalk cafés along the Côte d'Azur, our Gallic friends helped spread the word to the world. And despite a fall from grace and the much-maligned white

Zin of the 1970s, rosé recovered and became the "summer water" of the Hamptons elite in a Gatsby-esque twenty-first-century renaissance.

Take a dip into this fact-filled book and discover the fascinating history of the pink drink, along with truly informative details on the winemaking process, types of grapes, growing regions of the world and the best ways to select and serve your rosé. Celebrate the halcyon days of summer all year round with 20 cocktail recipes and a variety of food pairings. Amuse yourself with entertaining quotations and sayings that capture the love of rosé and explore fun facts and stats. There is something here for everyone, whether you enjoy a rosé mojito (a roséito) or you are a brosé who loves frozé.

After all, as one popular saying goes, "You can't buy happiness, but you can buy rosé, and that's close enough".

CHAPTER
ONE

The Rosé Revolution

Take a trip through the tumultuous history of rosé, from its humble beginnings in ancient Greece to today's resurgence – or "roséagence".

The first documented rosé dates back to about

7,000 BCE

Due to the method of producing wine, pink wine was the only one produced.

Although the Greeks and the Romans eventually figured out how to produce red and white wines, it wasn't until the sixth century BCE that the Phocaeans brought grape vines to Massilia in the south of France, and pink rosé started to become widespread.

According to legend, a violet-coloured rosé was created in Bordeaux during the Middle Ages. The wine picked up the nickname

"claret"

(in Latin, *claritas* means "clarity") and soon became fashionable around France.

In the Middle Ages, long after the fall of the Roman Empire, monastic orders continued to produce rosé, a staple in Provence, as a key revenue source for local abbeys.

The Medieval light
pink "claret" was treasured
as the finest wine available
and the darker pink-to-red
varieties were considered to
be of poor quality.

To the powerful English market, the most prized clarets were the

vin d'une nuit

or "wine of one night", made from juice that was allowed only a single night of contact with the grape skins.

A description of
rosé as

oeil de perdrix,

which in French translates to
"eye of the partridge", dates
back to the Middle Ages in the
Champagne region of France.

The name was a reference
to the pale pink colour of the
eye of a partridge struggling
in death's grip.

The making of champagne
in France was instrumental to
the history of rosé.

Most early champagne was
light pink in colour, or a darker
pink if the maker added
elderberries. It wasn't until
decades later that winemakers
developed a method for creating
white sparkling wines.

The classic Provençal rosé bottle is clear and curvy, a little like a bowling pin (skittle) or a corset and is sometimes referred to as a *flûte à corset*. In the 1930s, Marcel Ott, founder of Domaines Ott, and his son René created and patented it as their signature bottle.

When France introduced
mandatory vacation days
in the 1930s, families
flocked to Provence and
the Loire Valley where they
discovered rosé – it's been
associated with relaxing and
celebrating ever since.

Mateus, the famous Portuguese rosé, was created for an international market in 1942. The light, fizzy wine with flavours of watermelon and medicinal herbs was an easy-drinking alternative to beer, soda or champagne, and it soon became a household name.

The first rosé, Las Lanzas,
was imported to the US in the
mid 1940s.

In 1944 wine merchant
Henry Behar set sail for Portugal
to meet Maria da Fonsceca, the
maker of a rosé called Fascia,
which they renamed Las Lanzas
after Behar's favourite
Velasquez painting.

In France, rosé became *vin de soif,* a "wine to quench thirst" – a simple wine to drink while cooking or to offer as an aperitif before dinner. Many parents would even serve it, highly diluted, to their children.

After the First World War,
sweeter, sparkling rosés
gained in prominence
over the traditional drier
versions, leading the public
to mistakenly associate
rosé with sweet wine.

The first American-made rosé was created by Sutter Home in the 1970s when winemaker Bob Trinchero set aside his first pressing of grapes while attempting to create a more intense red Zinfandel.

Instead of discarding this "free run", he put it in his wine-tasting room, calling it white Zinfandel.

Later, through an accidental fermenting process, he created a sweeter version that became a phenomenal success.

By 1987, Sutter Home's white Zinfandel was the best-selling premium of all wines in the United States.

Although superseded by better-quality wines, it still enjoys a reputation as a good "gateway" wine.

To revive the Mateus brand in the 1970s, striking advertising campaigns featuring everyone from Jimi Hendrix to the Queen of England ran in the UK and were seen worldwide. The wine was instantly back in fashion.

Lancers rosé was introduced in the 1970s alongside Mateus.

Equally sweet and inexpensive, it was bottled in ceramic, with the unfortunate result that the wine would turn brown from oxidation.

The Mateus bottle
was inspired by the canteens
used by soldiers in the
First World War. And during
the Cold War period, Mateus
wine became one of the
most drunk wines by
US Army soldiers stationed
at European bases.

How "Blush" got its name

In 1976, wine writer
Jerry D. Mead visited Charles
Kreck at Mill Creek Vineyards in
Sonoma County, California, who
offered him a pale pink wine made
from Cabernet. Mead suggested
the name "Cabernet Blush".

In 1978 Kreck trademarked
the word, which became defined
for semi-sweet wines.

Using Pinor Noir grapes,
Tony Soter of Napa's Etude
introduced a dry rosé in 1992,
sparking a California trend
and putting rosé back on the map.

Roségence:
the
resurgence of rosé

❧

The revitalized interest in
the pink drink started at the
turn of the twenty-first century
and was aided by the popularity
of rosé as the summer drink in US
resorts, such as the Hamptons,
Miami and Beverly Hills.

Since the early 1990s, Long Island distinguished itself as a region for rosé, often producing dry rosés modelled on those made in southern France.

The eastern end of the island has over

60 VINEYARDS & WINERIES

that produce a range of rosés.

The first rosé produced on Long Island in 1992, in a batch of just 82 cases, was by Wölffer Estate – by 2015 they were turning out almost

22,000 CASES

for the summer season and the drink had become known as "Hamptons Gatorade".

In 2016 a New York Italian
bar called Bar Primi made
a rosé slushy with rosé
wine, vermouth and pureed
strawberries, called

the frosé,

that became a real-life
and Instagram hit. Some
versions substitute vodka
for the vermouth.

In 2017,

37 PER CENT

of the US adult population
were drinking rosé wine,
compared with

24 PER CENT

in 2007.

France – where rosé outstrips white wine in sales – consumes around

34 PER CENT

of the world's rosé wines and the US consumes

20 PER CENT

(surging by 43 per cent in one year alone).

Global consumption of rosé wine has soared in 17 years, growing an incredible

40 PER CENT

between 2002 and 2018.

CHAPTER
TWO

Where
there's a *will*,
there's a *Rosé*

Discover how this delicious
wine is made, the regions of
the world renowned for it
and how to choose your wine.
Get to know your grapes,
colours and flavour profiles.

In France it's called
ROSÉ,
in Italy it's called
ROSATO
and in Spain it's
ROSADO.

Rosé has been given the nickname "summer water" due to its refreshing taste and popularity over the summer months.

Blush or Rosé?

Although often used interchangeably, blush refers to the sweet white Zinfandel Californian grape, popular in the 1980s, which is jammy, fruity and fairly sweet.

Rosé normally is dry, fresh and has delicate aromatics and flavours.

Rosé wines are typically
less expensive than
red wines, because they are
simpler to make and don't
have to be aged.
So you don't need to spend
the big bucks to get
a fantastic-tasting rosé.

Rosé Won't Improve with Age

You might, but your rosé won't! As rosé spends little time in grape skins and is aged in stainless steel, it does not develop tannins or a deep colour.

Check the label and be sure to drink your bottle within two years (or sooner)!

Rosé has a reputation as a summer drink not just because it's most commonly served chilled and with lighter meals, but also because new batches are released in the spring and they tend to sell out by autumn.

Dry or Sweet?

Contrary to common belief, you can't really tell whether a rosé is dry or sweet from the colour. Traditionally, dry rosés are made from Grenache, Syrah, Sangiovese or Pinot Noir grapes, while sweet versions are made from Zinfandel, Merlot or Moscato.

In France, the drier rosés
come from the Loire Valley,
and have flavours of grapefruit
and mint.

In Bordeaux, rosé made
from Merlot is sweeter, with
aromatics of strawberry
and peach.

Contrary to popular thought, rosé is not a blend of red and white grapes. The pink colour is due to the production method, maceration or saignée, both of which use red grapes.

In Europe, rosé production is strictly regulated and the only acceptable reason for blending red and white grapes is to create pink champagne, made from Chardonnay, Pinot Noir and Pinot Meunier grapes.

Maceration

This is the most common method of making rosé. Red grapes are used, but they aren't kept in their skins and in contact with the juice for very long – in fact, anywhere from just 2 to 24 hours. The grapes are then squeezed for their juice and the skins discarded. The longer the skins sit in the juice, the stronger the colour, flavour, aroma and tannin.

Saignée

Translating as "to bleed", this method produces the most deeply pigmented and structured rosés. It begins with making a traditional red, but during fermenting, some of the juice is siphoned off to create the rosé. The result is that you can get both a rosé and a red out of one batch.

Grey Wine

Also called *vin gris*, this is the process by which a very pale rosé is produced.

The grapes are lightly pressed and produce a pale, almost grey, hue and a very delicate flavour.

The Charmat Method

Some producers are now using this technique for sparkling rosés, which carries out a secondary fermentation (which creates the bubbles) in a pressurized tank rather than in the bottles themselves.

The Charmat method originates from Italy where it is used to make Asti Spumante and prosecco.

How Pink is Pink?

Rosé comes in a vast range of colours from pale salmon to ruby. Common colour profiles include redcurrant, cantaloupe, peach and mango.

Tasting Profiles

You will most often find
one of these flavour
descriptors for your rosé:
grapefruit, raspberry, cherry,
strawberry or blackberry,
but also melon, orange,
celery, rhubarb or rose petal.

Pelure d'oignon

(onion skin) describes some slightly orange rosé colours. However never fear, the wines don't contain onions!

Wine referred to or labelled as "orange wine" is a white wine that finishes vinification on the skins – so even if it looks salmon-pink in colour, it is not a rosé.

> **"**
>
> *Rosé isn't just pink;*
> *it can be sunset*
> *cerise, weak ruby,*
> *subtle salmon or lurid*
> *bubblegum pink.*
>
> **"**

Suzy Atkins, *Telegraph.co.uk*

Centre de Rosé is a scientific research centre in the Var, France. Among its activities are tastings, training courses and the creation of a

10-FT (3-M)

Wall of Rosé displaying shelves of bottles in a dazzling array of pink shades.

When buying, don't simply
choose a rosé version
of your favourite red or
white brand.
Try small producers and
independents who
really dedicate themselves
to rosé.

Look for the term ABV (alcohol by volume) on the label, which is a great way to gauge whether the wine is sweet or dry. Wines with

12.5 PER CENT OR HIGHER

are typically dry, while those under that will most likely have residual sugar and will be off-dry or even sweet.

Don't judge a rosé by
its colour – the colour is
no indicator of the quality
of the wine.

Colour can be an indicator of flavour and style, however. Pale rosés tend to be lighter in body and more delicate, whereas darker varieties are drier with more body, robust flavour and a slightly higher alcohol content.

Fruity rosés can be made
with grapes such as Grenache,
Sangiovese, Mourvèdre or Pinot
Noir, while Cabernet Sauvignon and
Tavel varieties offer a more
savoury finish.

Colour Names for Rosé, from Light to Dark:

Mint	Wild strawberry
Grapefruit	Blood orange
Strawberry	Tomato
Tart cherry	Red bell pepper
Redcurrant	Blackcurrant
Sweet cherry	Blackberry
Raspberry	Berry jam

Get Down to the Grapes

Understand the basic grape types to pick your rosé.

Provence: Fruity and delicate, crisp and dry. Pale pink in colour with aromas of strawberry, watermelon and rose petal.

Grenache: Fruity with a brilliant ruby red colour and moderate acidity. Notes of ripe strawberry, orange and hibiscus.

Sangiovese: Fruity with a light-catching copper-red hue with good acidity. Notes of strawberry, plus green melon, roses and peach.

Tempranillo: Light style with a pale pink colour and herbaceous notes of green peppercorn, watermelon and strawberry.

Pinot Noir: Fruity and delicate, crisp and dry with bright acidity. Subtle notes of crabapple, watermelon, raspberry and strawberry.

Syrah: Savoury style in deeper ruby colours with notes of strawberry, cherry, peach and white pepper.

Cabernet Sauvignon: Savoury in a deep ruby red colour. Notes of deep cherry and blackcurrant.

Zinfandel: Sweet off-dry in pink hues with a moderately high acidity. Notes of strawberry, cotton candy, lemon and green melon.

Tavel: Savoury, rich and unusually dry with body and structure. A salmon-pink colour with notes of summer fruits.

Mourvèdre: Flowery and fruity with a fuller body and often pale coral in colour. Floral notes of violets and rose petals, with red plum, cherry and herbs on the palate.

The Most Famous Rosé

The rosés made in Provence in southeast France are incredibly popular and for good reason. They have a very pale colour, soft texture, "barely perceptible tannins" and are made from grape varieties that include Grenache, Cinsault and Mourvèdre, resulting in complex berry, citrus and floral notes.

When in doubt, buy Provence – they are the most consistent rosés and class leaders, taking away a slew of the top Global Master awards, year after year.

Grenache is the Grape

Many top-rated rosés are made with Grenache, or Garnacha as it is called in its native Spain, such as those from Gérard Bertrand, Château d'Esclans and Château Minuty. Look out for it blended with Vermentino (Rolle) grapes. It is a key player in the renowned Châteauneuf-du-Pape.

Rosé is Becoming a Luxury Wine

A barrel-fermented rosé from Gérard Bertrand in the 2020 Masters had characters of vanilla, toast, pineapple and citrus – and a retail of almost **£170 ($240).**

Provence is one of the
world's biggest rosé-
wine-producing regions,
dedicating

89 PER CENT

of their vines to rosé!

Tavel rosé, produced on the Rhône near Avignon, is the granddaddy of French rosé.

It's the first and the only appellation that exists today solely for rosé. The French call it a "gourmand" wine.

Rosé Regions to Visit – Besides Provence!

Tavel, Rhône Valley, France

Corsica, France

Navarra, Spain

Rioja, Spain

Puglia, Italy

Veneto, Italy

Sonoma and Napa Valleys, California

Long Island, New York

Top Rosé-producing Countries

France, 30 per cent

Spain, 21 per cent

USA, 14 per cent

Italy, 10 per cent

South Africa, 3 per cent

Germany, 2 per cent

Rosé d'Anjou from the Loire valley is a fresh off-dry rosé that became very popular in the twentieth century, when it accounted for

55 PER CENT

of the wine produced in the Anjou region.

Top-scoring rosés from
the US can be found
from states as diverse as
Washington, New York,
Virginia and Texas, as well as
California's Sonoma Valley.

66

A good rosé should be drier than Kool-Aid and sweeter than Amstel Light. It should be enlivened by a thin wire of acidity, to zap the taste buds, and it should have a middle core of fruit that is just pronounced enough to suggest the grape varietal (or varietals) from which it was made.

99

Jay McInerney, *Bacchus & Me: Adventures in the Wine Cellar*

66

Sweet wines aren't revitalizing on scorching hot days. Ice cold and darned dry is what you want.

99

Alan Richman, *GQ.com*

Angelina Jolie and Brad Pitt
launched a rosé wine in 2013
in collaboration with the Perrin
wine-making family from their
home Château Miraval, and
aptly named it Miraval.

The Perrin family also
created another rosé they
called Pink Floyd.

Whispering Angel claims
to be the best-selling rosé
in the world.

Produced by Château
d'Esclans, the wine has
gained a huge fan base
thanks to its refreshing,
crisp acidity.

CHAPTER
THREE

Rosé all Day

Whether you're
dining al fresco or
sipping poolside, rosé
has a place and time – here
are some ideas of how to
enjoy and serve your
rosé in style.

In summer in the
South of France, there's a saying
"Rosé All Day" to describe
having a glass of chilled rosé
with lunch or in the afternoon
on a lazy summer's day.

"

*On a shady terrace;
around a herb-scented
barbecue; outside a café
on market day; before
lunch by the pool – it
accompanies some of life's
most pleasant moments.*

"

Peter Mayle on rosé, *Provence A–Z*

66

*For me, summer is rosé
wine. And I love it.
It's the taste of summer.
I tend to prefer it
with a bit of club soda
in it, and ice.*

99

Nigella Lawson, NPR.org

66

*It is best enjoyed on an
afternoon walk through the
garden... If you're really warm,
it's okay to put an ice cube
in it. It extends the size of the
drink. Or freeze some of the rosé
and make a slush.*

99

Martha Stewart, *FoodandWine.com*

Like white wine, rosé is best served in a medium-sized glass so that the fresh and fruity characteristics gather towards the top.

You can serve rosé in a champagne flute, but choose a glass depending on the age of the wine.

For young, light rosés, a glass with a shallow bowl and a slightly flared lip places the wine on to the correct part of the palate for maximum enjoyment. Slightly more aged rosés may be served in white wine glasses.

The recommended serving
temperature for rosé is

7–13°C (45–55°F)

Be careful not to make it
so cold that you can't
taste anything.

Follow the 20:20 Rule.

Take your rosé or white out
the fridge 20 minutes before
serving (reds should go into
the fridge 20 minutes
before serving).

Don't Decant

Normally, decanting exposes oxygen to the wine, bringing out its best flavours, and is usually a good practice.

However, because rosé is so young, there is no need to decant.

Place rosé directly
into the refrigerator
after purchasing it, and
chill it for at least several
hours before serving (or
30 minutes in the freezer
if you're in a hurry).

To Ice or Not to Ice

Most sommeliers discourage
the addition of ice cubes to any
wine, since they will dilute and
change the colour, texture
and flavour as they melt. However,
many make an exception for
rosé, especially if the weather is
hot (or you happen to be in the
South of France!).

Rosé Piscine

Rosé or champagne served over ice is called La Piscine in France, which translates as "swimming pool". Movie star Brigitte Bardot coined the term after visiting Saint Tropez and, on being served the cocktail in a cognac snifter, remarked that the large glass looked like a swimming pool.

In France, rosé wine is drunk as an *aperitif*, when it's perfectly acceptable to serve it with ice.

In fact, it is even overtaking the anis-flavoured pastis in popularity, especially with younger female drinkers.

Another trend is the rosé *pamplemousse*: rosé wine blended with grapefruit (and sometimes peach) extract and sugar. This fairly low-alcohol, sweet drink can be bought in ready-mixed bottles in French supermarkets.

Rosé Après

A French tradition is to sip rosé après-ski – perhaps because it's sunny on the slopes and any sunny occasion merits rosé.

Women drink significantly more rosé than men in Germany, the Netherlands and the UK, but there doesn't seem to be a strong gender bias in France, the US, Russia or Canada.

Men drink more rosé than women in Brazil.

National Rosé Day

Held every second Saturday in June, National Rosé Day was founded in 2015 by Swedish winemaker Bodvár in order to "raise awareness and give rosé lovers a day to celebrate with 'summer's water' all over the world".

Provence rosé is generally agreed to set the gold standard for rosé, so it's not surprising that they also celebrate an International Rosé Day, every fourth Friday in June.

In Spain, rosado is enjoyed on grassy verges or streetside, in a practice called

botellón

translated as "pre-drinking". Everyone brings a bottle, glasses and some ice.

German rosé has almost doubled in production over the past decade – up to

12 PER CENT

of all German wine production. Their Pinot Noir plantings are the third highest in the world.

In Spain, you can find
a blend using red and white
grapes that sits between a rosé
and a red and is called

ojo de gallo

which means "eye of
the rooster".

Italian rosatos are diverse and the names can be perplexing.

A pink Lagrein wine is a *kretzer*, while on the southern shores of Lake Garda, a rosé is a *chiaretto*. In Abruzzo, it is called a *cerasuolo*, and at Carmignano in Tuscany, it is *vin ruspo*.

CHAPTER
FOUR

Stop and Smell the Rosé

Rosé is easy drinking.
You can just pop open your
plonk and go, but for
fine dining and finesse, here
are some fabulous ways
to serve it in cocktails or
pair it with food.

"

It is delightful, really. You can drink it with fish, chicken, broccoli, candy, Chinese food, cereal, chili, corn, hot dogs, rice, anything, out of a glass, bowl, pitcher, ladle, cupped hands, hat. It can be stored in the fridge, on a dining table, a counter, in a cabinet, an end table, really wherever you want!

"

Chrissy Teigen, about her husband John Legend's
LVE rosé, Instagram

"

*You can lob a cube
of ice in it without fear of
vitriol and serve it
with any food whatsoever.
It's a drink that says:
'I don't take life too
seriously.'*

"

Lorna Andrews, Instagram

"

Rosé wine is like your little black dress... there's one for breakfast, lunch and dinner.

"

Sommelier Belinda Chang, *Elle.com*

In 2020 Kylie Minogue released a moderately priced vegan rosé wine in association with Benchmark Drinks, which she said is "fresh, light and the perfect pink".

She suggests serving it with "baked salmon, topped with Japanese-style umami paste".

"

Rosés can be served with anything, but are usually reserved for cold dishes, pâtés, eggs and pork.

"

Julia Child, *Forbes.com*

66

Rosé is not only a great match for Mediterranean, but also for Japanese and other Asian cuisines, as well as Moroccan, Mexican and even Indian.

99

Jeany Cronk, *MirabeauWine.com*

Wine and Food Pairings

Light dry rosé, such as a Provençal, bardolino, chiaretto or a Pinot-based variety: Salads, goats' cheese, grilled fish and seafood.

Light off-dry rosé with a hint of sweetness, such as Rosé d'Anjou or Mateus Rosé: As for the light dry rosé, but best with salads and mild rice and pasta dishes.

Medium dry rosé and blushes, such as white Zinfandel: Salads, fish and light meals, but also great with lightly spiced foods and as a dessert wine with berries or melon.

Medium-bodied dry rosé, such as Rioja and Navarra: Versatile and suited for big flavours, such as pâtés, bruschetta, grilled meat and charcuterie, and recipes that include anchovies, olives and garlic.

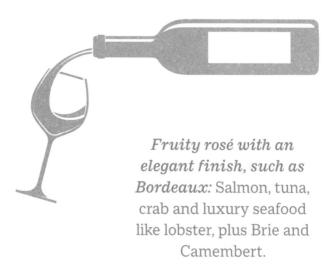

Fruity rosé with an elegant finish, such as Bordeaux: Salmon, tuna, crab and luxury seafood like lobster, plus Brie and Camembert.

Full-bodied fruity rosé, such as Syrah and Cabernet: With the mouth feel of a red, these are good with barbecues and spicy food.

Sparkling, including cava and champagne: All are excellent served with salads and light meals, but dry versions are best with canapés, and sweeter ones with cakes and pastries. Reserve champagne for lobster or rare meat and game.

PIMP YOUR ROSÉ

Five ways to take your drink
to another level.

1. Insert a sprig of fresh lavender
or rosemary for a herbaceous
aroma and taste.

2. Add ice - try small cubes or crushed
for a granita effect.

3. Drop in berries to enrich the natural
berry flavours.

4. Top with soda water or sparkling
water to add fizz.

5. Add a splash of vodka or gin to
dry and off-dry rosés.

LAVENDER ROSÉ COCKTAIL

Serves 1

2 fl oz (60 ml) rosé wine

1 tbsp (15 ml) grapefruit juice

1 tbsp (15 ml) lavender sugar syrup,
such as Monin

1 egg white

2½ fl oz (75 ml) prosecco or champagne

sprig of lavender, to garnish

Fill a cocktail shaker with ice and add the rosé, grapefruit juice, lavender sugar syrup and egg white. Shake vigorously until well chilled and strain into a chilled champagne flute or coupé-style glass. Top up with prosecco and add a sprig of lavender to garnish.

STRAWBERRY FROSÉ

Serves 4

1 bottle (750 ml) full-bodied rosé wine

½ cup (100 g) caster (superfine) sugar

8 oz (225 g) strawberries, hulled and quartered

juice of 1 lemon

extra strawberries, to garnish

To make the slushy, pour the wine into a baking tin (pan) and freeze 6 hours or overnight.

The next day, mix the sugar and strawberries together and leave to macerate for 30 minutes until the strawberries start to release their juices. Combine all the ingredients, except the garnish, together in a blender. Divide among four glasses, garnish with strawberries.

QUICK WATERMELON VODKA FROSÉ

Serves 4

8 fl oz (240 ml) rosé wine

4 fl oz (120 ml) vodka

4 fl oz (120 ml) watermelon juice

watermelon wedges and mint, to garnish

Combine all the ingredients, except the garnish, in a blender with plenty of ice. Divide among four glasses. Garnish with the watermelon and mint.

PINK FRENCH 75

Serves 1

1 fl oz (30 ml) gin

1 tbsp (15 ml) lemon juice

1 tbsp (15 ml) simple sugar syrup

sparkling rosé, to top

lemon twist, to garnish

Fill a cocktail shaker with ice and add the gin, lemon juice and sugar syrup. Shake vigorously, then strain into a chilled flute. Top with the sparkling rosé and garnish with the lemon twist.

Simple sugar syrup: Gently heat equal quantities of caster (superfine) sugar and water until the sugar dissolves. Allow to cool. Keep covered and stored in the fridge for up to two weeks. Variation: substitute rosé for the water.

ROSÉ SPRITZER

Serves 4

14 fl oz (400 ml) dry rosé wine

1 tablespoon grenadine

8 fl oz (240 ml) soda water

fresh strawberries and mint, to garnish

Mix the wine and grenadine in a pitcher and refrigerate until just before serving.

Pour into four individual glasses, top with the soda water and garnish.

ROSÉITO (ROSÉ MOJITO)

Serves 1

1 tbsp granulated sugar

juice of 1 lime

10 large fresh mint leaves

1¾ fl oz (50 ml) white rum

3½ fl oz (100 ml) rosé wine, such as
Côtes de Provence

soda water

mint sprig and edible rose petals, to garnish

Muddle the sugar, lime and mint in a glass
until the mint is broken down. Fill the
glass with ice, then add the rum and rosé.
Stir gently, top with soda water and serve
garnished with the mint and rose petals.

ROSÉ MARGARITA

Serves 1

1 fl oz (30 ml) lime juice, plus extra for rim

Himalayan pink salt, to garnish

1 fl oz (30 ml) silver tequila

1 fl oz (30 ml) simple sugar syrup
(*see page 122*)

4 fl oz (120 ml) rosé wine, such as
Côtes de Provence

Dip the rim of a chilled margarita glass into lime juice, then into the salt.

Fill a cocktail shaker with ice and add the lime juice, tequila and sugar syrup and shake vigorously until well chilled. Strain into the glass, then stir in the rosé.

ROSÉ SWAPS AND HACKS

If the party has finished, or you're coming to the end of the season, and you still have a reservoir of rosé left over, here are some ways to use it up.

1. Substitute fruity rosé for water when making simple sugar syrup and use it in cocktails.

2. Freeze the wine in ice-cube trays, with a leaf of sage, basil or mint in each, or a rosebud, to add a pretty touch to those last chilled drinks of the summer.

3. Add ¼ cup of a dry acidic rosé to a basic vinaigrette recipe of oil, red wine vinegar, honey and mustard. This goes especially well with salads containing citrus, cucumber or feta.

4. Use a dry rosé instead of white or red wine to deglaze meat, fish or seafood, but avoid dishes that are heavy on the cream or butter - this wine is best for lighter fare.

5. Combine a dry variety such as a Provence or Pinot Noir with vermouth and substitute it for red wine when cooking meat for heavier dishes like beef stew.

6. Include it in sauces, such as an orange sauce for duck or a pink peppercorn sauce for chicken.

7. Swap it for white wine when making risotto - rosé works especially well with mushroom or seafood versions.

8. A sweeter rosé is a great addition in desserts. Try using it for poaching pears or macerating berries.

ROSÉ SANGRIA

Makes 1 pitcher

1 bottle (750 ml) rosé wine,
preferably Spanish

3 peaches, stoned and sliced

7 oz (200 g) strawberries, hulled and sliced

1 tbsp caster (superfine) sugar

1¾ fl oz (50 ml) Grand Marnier

10 fl oz (300 ml) soda water

mint sprigs, to garnish

Put all the ingredients, except the soda
water and mint, in a large jug and stir to
combine. Refrigerate for 1 to 3 hours, until
chilled, then top up with soda water and
garnish with mint leaves.

RASPBERRY ROSÉ PUNCH

Makes 1 pitcher

1 bottle (750 ml) rosé wine

8 fl oz (240 ml) vodka

10 fl oz (300 ml) apple juice

1 fl oz (30 ml) raspberry sugar syrup,
such as Monin

7 oz (200 g) raspberries

Sparkling lemonade, to taste

1 lime, sliced

Put all the ingredients, except the sparkling
lemonade and lime, in a large jug and stir
to combine. Refrigerate for 1 to 3 hours,
until chilled, then top up with the sparkling
lemonade and garnish with lime slices.

BLACKBERRY ROSÉ COOLER

Serves 1

3 blackberries

1 tbsp (15 ml) lemon juice

1 tbsp (15 ml) vodka

1 tbsp (15 ml) simple sugar syrup

(*see page 122*)

3½ fl oz (100 ml) rosé wine

blackberry, to garnish

In a cocktail shaker, muddle the blackberries with the lemon juice, vodka and sugar syrup, until the blackberries have broken down. Fill the shaker with ice, shake vigorously until well chilled, and strain into a highball glass. Stir in the rosé wine and garnish with a blackberry.

ROSÉ ELDERFLOWER SPRITZER

Serves 1

4 fl oz (120 ml) dry rosé wine

1 fl oz (30 ml) elderflower liqueur, such as St Germain's

1 tbsp (15 ml) fresh lemon juice

soda water

lemon slice, to garnish

Gently mix the rosé, elderflower and lemon juice together, then pour over ice in a glass. Top with soda water and garnish with the lemon slice.

ROSY MARTINI

Serves 1

2½ fl oz (75 ml) rosé wine

1 fl oz (25 ml) dry vermouth

fresh edible rose, to garnish

Add the alcohol to a cocktail shaker with
ice and shake vigorously until well chilled.
Strain into a chilled martini glass and garnish
with the edible flower.

ROSY GIMLET

Serves 1

2½ fl oz (75 ml) rosé wine

1 fl oz (25 ml) gin

1 fl oz (25 ml) lime juice

lime twist, to garnish

Add the alcohol to a cocktail shaker with ice
and shake vigorously until well chilled. Strain
into a chilled martini glass and garnish
with the lime.

APEROL ROSÉ SPRITZ

Serves 1

½ orange

2 fl oz (60 ml) Aperol

3 fl oz (90 ml) sparkling rosé wine

soda water

orange slice, to serve

Cut a slice from the orange for the garnish and squeeze the juice from the rest. Fill a glass with ice and pour in the orange juice, Aperol and sparkling rosé. Stir gently, then top with soda water. Garnish with the orange slice.

KIR ROSÉ

Serves 1

2 tsp crème de cassis or a black raspberry
liqueur, such as Chambourd

6 fl oz (180 ml) sparkling rosé wine

Pour the cassis or Chambourd into a
chilled flute, top with the sparkling
rosé and stir gently.

SPARKLING GIN AND GRAPEFRUIT COCKTAIL

Serves 1

1 fl oz (30 ml) gin

1 fl oz (30 ml) freshly squeezed pink grapefruit juice

2 fl oz (60 ml) sparkling rosé wine

sprig of rosemary, to garnish

Fill a cocktail shaker with ice, add the gin and grapefruit juice and shake vigorously until well chilled. Strain into a chilled glass, top with the sparkling rosé and serve garnished with the rosemary.

LEMON POMEGRANATE ROSÉ

Serves 1

6 fl oz (180 ml) rosé wine

1 tbsp simple sugar syrup (*see page 122*)

2 tbsp pomegranate juice

splash of sparkling lemonade or lemon water

1 tbsp fresh pomegranate seeds

lemon slice, to garnish

Mix the rosé, sugar syrup and pomegranate juice in a glass with ice. Top with a splash of sparkling water. Drop in the pomegranate seeds and serve garnished with the lemon.

ROSÉ COSMO

Serves 1

1 fl oz (25 ml) vodka

1¾ fl oz (50 ml) cranberry juice

juice from ½ lime

2½ fl oz (75 ml) rosé wine

Fill a cocktail shaker with ice. Add all the ingredients and shake vigorously until well chilled. Strain into a chilled coupé-style glass.

ROSÉ SOUR

Serves 1

2 fl oz (60 ml) Bourbon whiskey

1 tbsp (15 ml) freshly squeezed lemon juice

1 tbsp (15 ml) simple sugar syrup
(*see page 122*)

1 fl oz (25 ml) dry rosé

Fill a cocktail shaker with ice. Pour in the
whiskey, lemon juice and sugar syrup. Shake
vigorously until well chilled. Strain into an
ice-filled short glass, then slowly pour the rosé
over the back of a spoon to float on top.

CHAPTER
FIVE

Through Rosé-coloured Glasses

Join the celebrities and influencers extolling the virtues of rosé on Instagram and Pinterest with these inspiring quotes on living your best rosé life.

66

Life isn't all diamonds and rosé... but it should be.

99

Lisa Vanderpump, *TheZoeReport.com*

66

I think a rosé should have that inherently Pavlovian-to-women peachy-pink quality that just draws us in. I don't know what it is about us girls, but we love pink.

99

**Drew Barrymore, on her own-brand
Carmel Road rosé, Instagram**

66

Rosé is kind of like online dating. What was once a faux pas has become the norm. It's totally become universally accepted among men and women. It's kind of like the beer of wine.

99

Sam Daly, *GQ.com*

❝

Is it time for a rosé yet?

❞

Nadia Fairfax, Instagram

66

Pink isn't just a colour. It's an attitude.

99

Miley Cyrus, *EliteDaily.com*

❝

Rosé is for when you want to get a little fancy.

❞

Post Malone, *NYPost.com*

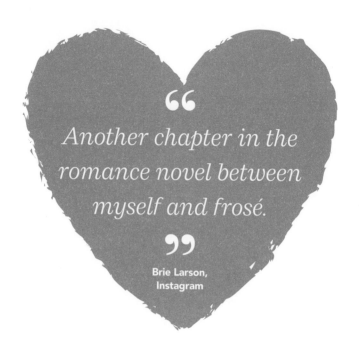

66

Another chapter in the romance novel between myself and frosé.

99

**Brie Larson,
Instagram**

"

Rosé.

"

Emma Stone's reply, when asked whether she
preferred red or white, for *The Hollywood
Reporter*'s annual beauty issue

66

I love rosé more than any member of my biological family.

99

Josh Ostrovsky, who set up Swish to create White Girl Rosé, Instagram

66

Happy July 4th. I'm crushing cans of rosé and eating smoked meat and doing nude cannonballs, which is exactly how George Washington would have wanted it right?

99

Josh Ostrovsky, Instagram

"

Alright, I know,
I know, give me some
Hamptons juice.

"

Anthony Bourdain, on being caught buying
several bottles of rosé in the Hamptons, *Refinery29.com*

"

My dad had always referred to rosé as 'pink juice', but that night, we said, 'No, no, no. We call it Hampton Water now'.

"

Jesse Bongiovi, on the rosé Diving Into Hampton Water he created with his father Jon Bon Jovi, *CNBC.com*

"

I'm drinking a glass of 'lady petrol'.

"

Jeremy Clarkson drinks rosé, Twitter

"

There's no snobbishness to it. It's just something you drink because you want to feel happy.

"

Jeremy Clarkson on rosé, *The Sun*

"Yes way, rosé."

"Rosé all day."

"Where there's a will, there's a rosé."

"Rosé the day away."

*"Stop and smell
the rosé."*

*"Rosé to the
occasion."*

Unknown, as seen on Pinterest

"

Rosé instantly inspires these good, happy, positive vibes.

"

Erica Blumenthal of Yes Way Rosé, *Thrillist.com*

You can drink it at the pool out of a red Dixie cup, and you don't have to break out your Riedel sommelier series wine glasses.

Sommelier Belinda Chang, *Thrillist.com*

66

*Someone give me
a book and rosé.
I'll do the rest.*

99

Ashley Tisdale, Instagram

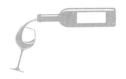

66

Rosé is like coffee to my brain. It just makes me feel more alive.

99

Máté Schubert, *IntotheRose.com*

66

It's like a waking-up wine, a get out of bed and drink wine... It's good for while you're playing tennis – unless you're a professional!

99

David Matthews on his Dreaming Tree rosé,
***People* magazine**

66

Um guys... They have steins of rosè in Germany.

99

Jerry O'Connell, Instagram

"

I put rosé in my Bolognese. Making dinner with love.

"

Lady Gaga, Instagram

"

Doing my patriotic best.

"

Sarah Michelle Geller, drinking rosé on the
4th of July, Instagram

"You had me at rosé."

"Slay, then rosé."

"You can't buy happiness, but you can buy rosé, and that's close enough."

"La vie en rosé."

"I'll take rosé over roses today, thank you."

"The world looks better through rosé-coloured glasses."

Unknown, as seen on Pinterest

66

I prefer my wine sparkling, pink and under $11.

99

Jess Day, in *New Girl*, Season 5, Episode 12

66

*Pink wine makes
me slutty.*

99

Jess Day, in *New Girl*, Season 1, Episode 1

66

I blame the rosé!!

99

Ellen Pompeo, on having her purse stolen in Italy, *MSN.com*

66

Press day.
Frozé! Tuesday!
I hate myself.

99

Brittany Snow, drinking frozé on Instagram

"

So we have a few options: there's the strawberry-peach, the ground fruits blend and... a lovely banana rosé.

"

Herb describes the three contenders for Moira Rosé,
Schitt's Creek, Season 6, Episode 7

"

I've been known to sample the occasional rosé. And a couple summers back, I tried a Merlot that used to be a Chardonnay, which got a bit complicated.

"

David to Stevie, *Schitt's Creek*, Season 1, Episode 10

"I was told you'd have rosé."

"Fifty shades of rosé."

"Anything is possible with sunshine and a glass of rosé."

*"Can't pay rent.
Can buy rosé."*

*"Today's Forecast:
99 per cent chance
of rosé all day."*

"It's raining rosé."

Unknown, as seen on Pinterest

CHAPTER
SIX

In the
Pink

The meteoric rise of rosé has put this pink in the pink. Here are facts and stats behind the drink's boost in popularity, as well as some surprising health benefits.

66

If you're cutting down on sugar, it's good to know that rosé wine usually has less sugar than red or white.

99

Kourtney Kardashian, *People.com*

A glass of rosé wine
contains around

82 CALORIES,

making it one of the lowest
calorie alcoholic drinks.

Yes, There are Health Benefits!

Rosé wine contains many useful compounds with potential anti-inflammatory properties and antioxidants.

1. Polyphenols help lower LDL (or bad) cholesterol levels.

2. Potassium balances metabolism and blood pressure, and regulates water in the body.

3. Resveratrol is the anti-ageing antioxidant that also protects against cancer.

4. Anti-inflammatory properties help to lower the chances of rheumatoid arthritis.

5. Darker varieties have more anthocyanins, the colourants in red grapes associated with protection against coronary heart disease.

Choose the Pinot Noir
grape, which has a higher
resveratrol concentration
than any other red grape.

❧

Studies have also suggested
it can improve brain health
and insulin sensitivity.

Research has shown
white wine drinkers have a

13

PER CENT

higher risk of cancer than
red or rosé drinkers.

All-natural ingredients,
low sugar and low ABV,
Rose Rosé by Haus is
marketed for those concerned
about health.

It's made with Chardonnay
grapes, elderflower
and lemon and boasts
notes of raspberry, rose and
watermelon.

Sugar content in wine is generally measured by the levels of residual sugar left over from grapes after the fermentation process, and sweet rosés can have between

35 AND 120 GRAMS.

Choose dry!

To cut on calories and the
amount you drink,
serve your rosé in small
glasses – rosé doesn't need
to breathe.

According to Nielsen,
40 PER CENT
of rosé wine consumers
are females aged
21 TO 34
who are most likely to
scroll through social
media. It is most popular
in coastal, cosmopolitan
and urban settings.

A few popular brands – such as White Girl Rosé and Yes Way Rosé – owe their inception to Instagram's promotion of rosé as a fun "lifestyle" drink, and of course it comes in millennial pink.

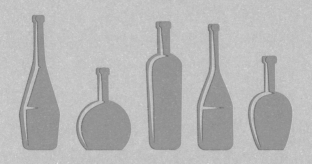

One in every ten bottles
of wine sold in the UK is pink.

Over the summer it
increases to more than
one in eight.

An explosion of celebrity
rosés have launched
on to the market in recent
years, from Sir Ian Botham
and Jon Bon Jovi to
John Malkovich and Sarah
Jessica Parker.

The popularity of rosé has grown among men too, leading to the term "brosé" to describe men who "pound the pink".

The brosé effect has spawned many social-media postings as well as seemingly serious discussions on the liberating effect of confessing your love for the drink.

"

I like to say that real men drink pink.

"

Thomas Pastuszak, wine director, *GQ.com*

Nearly
3 MILLION
pictures on Instagram are
tagged #rosé, showing
appealing images of
people enjoying the
"Rosé All Day"
lifestyle.